s. Wallis

RUTGERS UNIVERSITY PRESS
New Brunswick, New Jersey, and London

Revised edition with new cartoons

EINSTEIN SIMPLIFIED

Cartoons on Science
by Sidney Harris

Revised edition, 2004

Library of Congress Cataloging-in-Publication Data

Harris, Sidney.
  Einstein simplified : cartoons on science / by Sidney
Harris.
    p. cm.
  ISBN 0-8135-3386-4 (pbk.)
  1. Science—Caricatures and cartoons. 2. American
wit and humor, Pictorial. I. Title.
NC1429.H33315A4 1989
741.5'973—dc19   88-31284
                                                   CIP

British Cataloging-in-Publication information available

All the cartoons in this book have been previously published
and copyrighted by the following periodicals: *American
Scientist, Bulletin of the Atomic Scientists, Chicago Magazine,
Clinical Chemistry News, Discover, Environment, Fantasy
and Science Fiction, Hippocrates, Johns Hopkins Magazine,
Phi Delta Kappan, The New Yorker, Physics Today, Playboy,
Punch, Saturday Review, Science, Science 80* (and *Science
81, 83, 84,* and *85*). The cartoons have been reprinted with
permission, and our thanks go to all the above publications.

Cartoons on pages 60, 109, 142, 148, copyright © 1970,
1982, 1986, 1971 *Playboy.*

Cartoons on pages 17, 37, 92, 118 and 119, 150, copyright
© 1986, 1973, 1983, 1978, 1979 *The New Yorker.*

Manufactured in the United States of America

Einstein Simplified

"THERE GOES ARCHIMEDES WITH HIS CONFOUNDED LEVER AGAIN"

"JUST CHECKING."

"WE COLLABORATE. I'M AN EXPERT, BUT NOT AN AUTHORITY, AND DR. GELBIS IS AN AUTHORITY, BUT NOT AN EXPERT."

"IT LOOKS LIKE FRED HAS GONE SOLAR."

MESOZOIC ERA ➤ ◄ CENOZOIC ERA

"IT FIGURES. IF THERE'S ARTIFICIAL INTELLIGENCE, THERE'S BOUND TO BE SOME ARTIFICIAL STUPIDITY."

RAIN FOREST

EMERGENTS →

CANOPY →

CLIMBERS →

UNDERSTORY →

UMBRELLA
SALESMEN →

S.Harris

"THE ENVIRONMENT PEOPLE ONLY WORRY ABOUT ENDANGERED SPECIES, NOT ENDANGERED INDIVIDUALS."

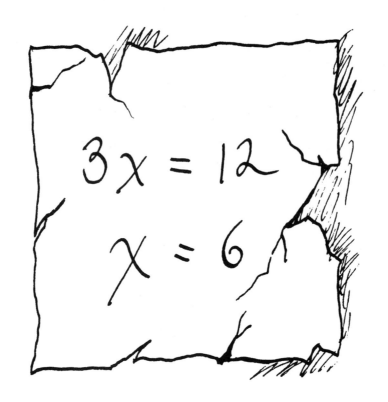

$$3x = 12$$
$$x = 6$$

EINSTEIN'S
FIRST EQUATION

"IF I LEARNED ONE THING IN MY LONG REIGN, IT'S THAT HEAT RISES."

THE LAST BRONTOSAURUS ON EARTH UNABLE
TO UNDERSTAND WHY HE CAN'T GET A DATE
FOR SATURDAY NIGHT

"SINCE THAT YOUNG DARWIN FELLOW WAS HERE, THEIR SPECIES CERTAINLY HAS EVOLVED."

"UNFORTUNATELY THIS LAB IS FUNDED ONLY BY AS MUCH GOLD AS WE CAN MAKE FROM LEAD."

"DON'T FORGET—KEEP THE POTASSIUM CHLORIDE IN A SEPARATE CONTAINER."

"So, by a vote of 8 to 2 we have decided to skip the industrial revolution completely, and go right into the electronic age."

"ALL RIGHT, SO HE DROPPED THE HEART. THE FLOOR IS CLEAN."

IMMEDIATELY AFTER ORVILLE WRIGHT'S HISTORIC 12-SECOND FLIGHT, HIS LUGGAGE COULD NOT BE LOCATED.

"THEN, AS YOU CAN SEE, WE GIVE THEM SOME MULTIPLE CHOICE TESTS."

"IT'S BETTER THAN WE EXPECTED. NOW THE SHEEP PRODUCE WOOL IN THE WINTER AND COTTON IN THE SUMMER."

"THE ONLY PART OF THE UNIVERSE WHICH ISN'T EXPANDING IS THE BUDGET FOR THIS PLACE."

TRYING TO DESCRIBE THE SIZE OF THE BIG BANG

"BIGGEST DAMN VIRUS I'VE EVER SEEN!"

HOW TO PLAY WITH A
RADIOACTIVE BASEBALL

GREATLY ENLARGED VIEW OF NEWLY DISCOVERED PARTICLE

"IT MAY NOT BE A PERFECT WHEEL, BUT IT'S A STATE-OF-THE-ART WHEEL."

TECHNOLOGY BREAKTHROUGH:
THE CORDLESS TOASTER

"DON'T KID YOURSELF. IF YOU'RE TOO UGLY TO BE COOKED, THEY'LL GRIND YOU UP INTO ONE OF THOSE FISH PROTEIN CONCENTRATES."

"I MISS THE GOOD OLD DAYS WHEN ALL WE HAD TO WORRY ABOUT WAS NOUNS AND VERBS."

AS SMART AS HE WAS, ALBERT EINSTEIN COULD NOT FIGURE OUT HOW TO HANDLE THOSE TRICKY BOUNCES AT THIRD BASE.

"THIS IS TERRIBLE. IF THE PLANKTON GO OUT ON STRIKE, IT'LL DISRUPT THE ENTIRE FOOD CHAIN."

INSTITUTE of
SOLID GEOMETRY

Institute of
Plane
Geometry

PEMBROOK, THE RENEGADE THIRD WRIGHT BROTHER, AT WORK ON THE FIRST ANTI-AIRCRAFT GUN

"ALL I CAN TELL YOU IS THEY'RE <u>NOT</u> THE CUTE LITTLE BIRDS WHO USUALLY RIDE AROUND ON US."

"IT STARTED WITH JUST THE PARTICLES BEING ACCELERATED, BUT NOW EVERYTHING AROUND HERE HAS SPEEDED UP."

"IF ONLY HE COULD THINK IN ABSTRACT TERMS."

"I ADMIRE THE INQUIRING MIND AND THE PRAGMATIC MIND, BUT I ALSO ADMIRE SOMEONE WHO CAN HIT."

"QUARKS. NEUTRINOS. MESONS. ALL THOSE DAMN PARTICLES YOU CAN'T SEE. THAT'S WHAT DROVE ME TO DRINK. BUT <u>NOW</u> I <u>CAN</u> <u>SEE</u> THEM!"

"OF COURSE IT'S SAFE. IT HAS NO PRESERVATIVES, NO ADDITIVES, NO ARTIFICIAL COLORING..."

A NATION OF EINSTEINS

1.

2.

S.harris

High
Bouncers
12 Lire

"A GIANT KILLER MACROPHAGE! DOCTOR, THIS IS MADNESS!"

MAMMALS BIRDS REPTILES FISH

EVOLUTION UPDATE — 101,989 A.D.

"RORSCHACH! WHAT'S TO BECOME OF YOU?"

"THE FUNNY THING IS, OUR WASTES ARE PERFECTLY HARMLESS."

I.Q. TEST FOR BIRDS

"IF WE ANNOUNCE WHAT WE'VE FOUND, IT'LL GET A SMALL ARTICLE IN THE PAPER. IF WE SAY WE CAN'T FIGURE OUT WHAT IT IS, IT'LL MAKE THE FRONT PAGE."

"PARTICLES, PARTICLES, PARTICLES."

BARTENDER WITH
P.H.D. IN PHYSICS

"HE SALIVATES."

"ALL WE WANT IS SOMETHING NEW THAT WILL INCAPACITATE THE ENEMY WITHOUT GIVING US A BAD PRESS."

SCIENCE AND SOCIETY~1923

SCHRÖDINGER, PAULI, HEISENBERG, PLANCK,
BOHR, CURIE, EINSTEIN, KIKI AND BUD VANDERVELT

"IT'S OUR NEW ASSEMBLY LINE. WHEN THE PERSON AT THE END OF THE LINE HAS AN IDEA, HE PUTS IT ON THE CONVEYOR BELT, AND AS IT PASSES EACH OF US, WE MULL IT OVER AND TRY TO ADD TO IT."

"LOOK—UP IN THE SKY! IT'S A FLOCK OF BIRDS,
IT'S A CLOUD. IT'S THE 'MONTGOLFIER BROTHERS."

EVEN AMONG THE SOCIAL INSECTS, SOME INDIVIDUALS ARE CLEARLY ANTI-SOCIAL

EINSTEIN SIMPLIFIED

s.harris

"GRANTED, WE HAVE TO DO THE RESEARCH. AND WE CAN DO SOME RESEARCH ON THE RESEARCH. BUT I DON'T THINK WE SHOULD GET INVOLVED IN RESEARCH ON RESEARCH ON RESEARCH."

"As I mentioned next week in my talk on reversible time..."

"...AND THIS, OF COURSE, IS THE CLEAN ROOM."

"MY GOODNESS, IT'S 12:15: 0936420175. TIME FOR LUNCH."

"LET'S GO OVER TO CELSIUS'S PLACE. I HEAR IT'S ONLY 36° OVER THERE."

"IN THE BEGINNING, BIOPHASE GENETICS, INC. CREATED US."

"THE RIGHT SIDE OF MY BRAIN SAYS YES, BUT I'M WAITING TO HEAR FROM THE LEFT SIDE OF MY BRAIN."

"I TRIED IT, BUT THE
BLOOD RUSHED TO MY FEET."

"THE BEAUTY OF THIS IS THAT IT IS ONLY OF THEORETICAL IMPORTANCE, AND THERE IS NO WAY IT CAN BE OF ANY PRACTICAL USE WHATSOEVER."

CONFERENCE ON THE CONTROL OF TIME AND SPACE
Left to Right: ISAAC NEWTON, ARISTOTLE, H.G. WELLS, PTOLEMY, ISAAC ASIMOV, ALBERT EINSTEIN, GALILEO, COPERNICUS

"I'LL BE WORKING ON THE LARGEST AND SMALLEST OBJECTS IN THE UNIVERSE — SUPERCLUSTERS AND NEUTRINOS. I'D LIKE YOU TO HANDLE EVERYTHING IN BETWEEN."

"WE TRIED HUNTING AND WE TRIED GATHERING, BUT NOW WE USUALLY EAT OUT."

1.

INGREDIENTS:
SUGAR, FLOUR,
SHORTENING, SALT,
ARTIFICIAL FLAVORING,
BHA and BHT,
PROPYL GALLATE,
SODIUM PROPIONATE,
MONO and
DIGLYCERIDES

DORVAL'S
CaKe
Mix

2.

3.

"IT SEEMS TO BE FULL OF DATA."

"DON'T MIND ASHLEY. AFTER LOOKING THROUGH A MICROSCOPE ALL DAY, ANYTHING LARGE STARTLES HIM."

"DR. HODGES, HERE, IS FROM ENGLAND, AND HE'S BEEN OBSERVING US FOR 14 YEARS. MR. FERRELL, AN AMERICAN, HAS BEEN HERE ONLY THREE WEEKS. MONIQUE CORVEAU, FROM PARIS, HAS PRACTICALLY BEEN LIVING WITH US FOR ABOUT NINE YEARS..."

"I HOPE THEY DON'T ISOLATE ME. THEY'LL NEVER BELIEVE I'M HARMLESS."

1.

2.

3.

4.

DARWIN

"HE HASN'T MADE ANY PROGRESS WITH HIS THEORIES RECENTLY, SO HE'S BEEN WORKING ON HIS RESEMBLANCE TO EINSTEIN."

AFTER ANTOINE LAVOISIER DISCOVERED
OXYGEN, HIS WHOLE FAMILY WOULD BREATHE IT
REGULARLY AS A SHOW OF SUPPORT

EARLY SCIENTIFIC FRAUD:
YOUNG THOMAS EDISON TRIED TO PASS OFF A
CONTAINER FILLED WITH FIREFLIES AS AN
INCANDESCENT BULB

PERSONAL AIR BAG

"IT STARTED WITH A SIMPLE CASE OF PEER-REVIEW."

"OUR REPUTATION FOR LONGEVITY IS BASED ON SEVERAL FACTORS: HARD WORK, SIMPLE FOOD, LACK OF STRESS, AND THE INABILITY TO COUNT CORRECTLY."

"WE HOPE TO MAKE ANTIBIOTICS, INTERFERON AND DIAGNOSTIC PRODUCTS, BUT JUST TO BE ON THE SAFE SIDE, WE'RE STARTING OUT WITH A LINE OF SHAMPOOS."

"EYE IRRITATION AND COUGHING? YOU CALL THAT TOXIC?"

"I THOUGHT HE WOULD RUN ALL SORTS OF SCIENTIFIC TESTS."

"ONE NICE THING ABOUT BEING A CEPHALOPOD —
YOU PUT ON YOUR SHOES, YOU PUT ON YOUR
HAT, AND YOU'RE DRESSED."

"WE DID THE WHOLE ROOM OVER IN FRACTALS."

"THEY DID IT AGAIN — NOT A WORD IN THE WEATHER REPORT ABOUT AN ICE AGE."

"WE CREATE IT, WE CLEAN IT UP— BUSINESS COULDN'T BE BETTER."

"WE'VE DISCOVERED A MASSIVE DUST AND GAS CLOUD WHICH IS EITHER THE BEGINNING OF A NEW STAR OR JUST A HELL OF A LOT OF DUST AND GAS."

"I THOUGHT THEY ONLY TESTED DRUGS ON GUINEA PIGS."

THE UNIVERSE BEFORE
THE BIG BANG
(ACTUAL SIZE)